STARK LIBRARY APR 2021

T4-ADM-007

A Question of Science

How Can a Plant Eat a Fly?

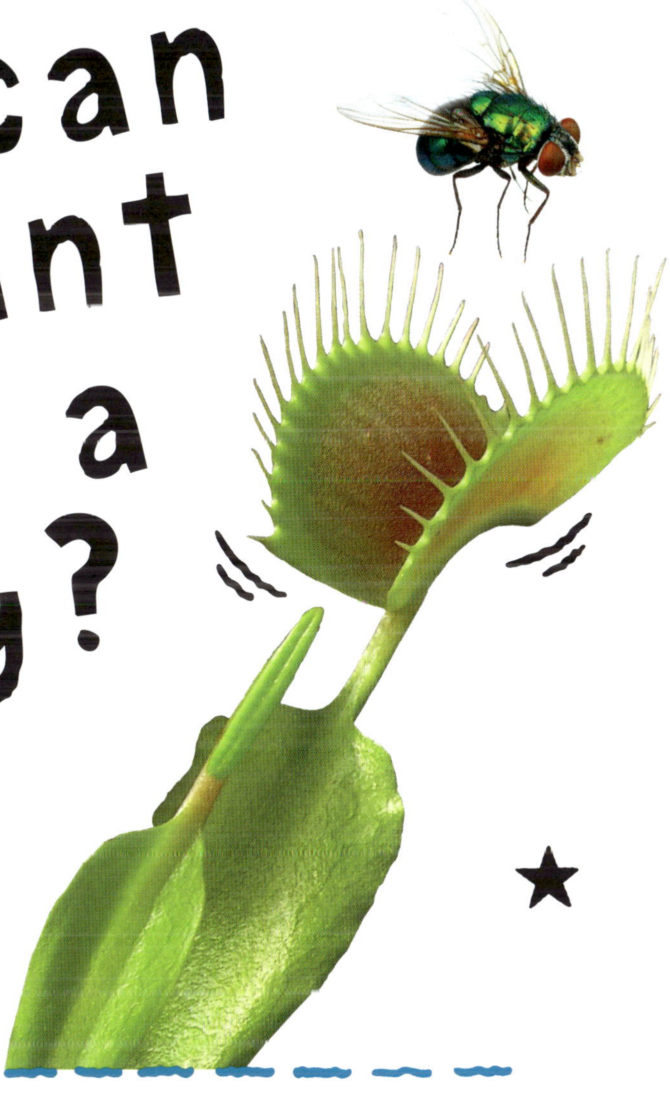

And other questions about PLANTS

Anna Claybourne

Published in Canada
Crabtree Publishing
616 Welland Ave.
St. Catharines, Ontario
L2M 5V6

Published in the United States
Crabtree Publishing
347 Fifth Avenue
Suite 1402–145
New York, NY 10016

Published in 2021 by Crabtree Publishing Company

All rights reserved. No part of this publication may be reproduced, stored in a retrieval system or be transmitted in any form or by any means, electronic, mechanical, photocopying, recording, or otherwise, without the prior written permission of the copyright owner.

First published in 2020 by Wayland
© Hodder and Stoughton 2020

Author: Anna Claybourne
Editorial Director: Kathy Middleton
Editor: Julia Bird
Proofreader: Petrice Custance
Design and illustration: Matt Lilly
Cover design: Matt Lilly
Production coordinator and Prepress technician: Tammy McGarr
Print coordinator: Katherine Berti

Picture credits
Alamy: Arco Images GmbH 8cr; Nigel Cattlin 18b; Valentyna Chukhlyebova 27c; Tim Gainey 14b; NASA 27b.
Dreamstime: Jose Luis Agudo 18t.
Getty Images: Peter Bennett 20bl; Andrew Lichtenstein/Sygma 21b;
Nature PL: Cyril Ruoso 9cl.
Shutterstock: Africa Studio 20t; Alexeysun 7cr; AP/REX 26b; Aprily 13cl; Artiste2d3d 5cl; Caron Badkin 17tr; Barbol 7c; Randy Bjorklund 17cr; Todd Boland front cover b,1b, 24cl; Radek Borovka 13b;Aleksandra Budzinskaia 9bl; Dan Campbell 29t; Cat Act Art 9tl; Dewib'Indew 5c; digidreamgrafix 13tr; digital reflextions 7b;Dotted Yeti 27t; Ekaterina43 19t; Emilio100 19b; fizkes 8br; Focal Point 23b; Kim Fooyontphanich 16tr; Volodymyr Goink 29br; Chris Hoff 18c; irin-k front cover, 1c, 9cl, 24t; Kaiskynet Studio 29bl; Sergey Kaykov 6cl; Ekaterina Kondratova 8bc; kpboonjit 5cr; LEOCHEN66 25; Madlen 5t; marich 7tl; Mazur Travel 10c, 15b; Dudarev Mikhail 13c; mikolajn 20bc; Mr Anuwat 10br; Theerasak Namkampa 9br; Nenov brothers Images 9cr; Anna Om 22t; Sari Oneal 15c; Palephotography 28br; Perutskyi Petro 8bl; Piyathep 9tr; Volodymyr Plysiuk 8cl; Wichai Prasomsri1 8cl; Quick shot 4; Valentina Razumova 10bc; Sue Robinson 22b; Roman Samokhin 10cr; SanderMeertinsPhotography 15t; sharpner 23c; showcake 10bl; Pairoj Sroyngem 28bl; Kuttlevarasova Stuchelova 24r; sutham 19cr; tektur 12; trappy76 20bcr; Ajay Tvm 19cl; Velasquez77 6cr; Simic Vojislav 17tcl; Matka_Wariatka 5ccr.
CC Wikimedia Commons: DASonnerfield: 16bl.

Printed in the U.S.A./082020/CG20200601

Library and Achives Canada Cataloguing in Publication

Title: How can a plant eat a fly? : and other questions about plants / Anna Claybourne.
Names: Claybourne, Anna, author.
Description: Series statement: A question of science | Includes index.
Identifiers: Canadiana (print) 20200254162 | Canadiana (ebook) 20200254189 |
ISBN 9780778777533 (softcover) |
ISBN 9780778777069 (hardcover) |
ISBN 9781427125385 (HTML)
Subjects: LCSH: Plants—Juvenile literature. | LCSH: Plants—Miscellanea—Juvenile literature. | LCGFT: Trivia and miscellanea.
Classification: LCC QK49 .C53 2020 | DDC j580—dc23

Library of Congress Cataloging-in-Publication Data

Names: Claybourne, Anna, author.
Title: How can a plant eat a fly? : and other questions about plants / Anna Claybourne.
Description: New York, NY : Crabtree Publishing Company, 2021. | Series: A question of science | First published in 2020 by Wayland.
Identifiers: LCCN 2020023615 (print) | LCCN 2020023616 (ebook) |
ISBN 9780778777069 (hardcover) |
ISBN 9780778777533 (paperback) |
ISBN 9781427125385 (ebook)
Subjects: LCSH: Plants--Juvenile literature.
Classification: LCC QK49 .C578 2021 (print) | LCC QK49 (ebook) | DDC 581--dc23
LC record available at https://lccn.loc.gov/2020023615
LC ebook record available at https://lccn.loc.gov/2020023616

Contents

What are plants?	4
Why don't plants have mouths?	6
Could we exist without plants?	8
Why can't plants walk around?	10
How do cacti survive in the desert?	12
Why do flowers smell nice?	14
How can a seed grow after thousands of years?	16
Do plants have feelings?	18
Why are trees so big?	20
Can plants talk to each other?	22
How can a plant eat a fly?	24
Are there plants in space?	26
Quick-fire questions	28
Glossary	30
Learning more	31
Index	32

What are plants?

If you were an alien landing on planet Earth, one of the first things you would notice is all the plants.

IT'S A BIT GREEN AROUND HERE!

Green planet

Even with all of its towns, cities, roads, ice, and rock, much of Earth is covered by plants. Forests make up about 30 percent of land surface. Plants are also found in parks, gardens, swamps, and grasslands. There are even more plants in rivers and oceans.

So what are plants?

A plant is a living thing that makes its own food. To do this, it needs three key ingredients:

- sunlight
- water
- **carbon dioxide** gas from the air

A typical plant has these main parts to keep it alive.

Tomato plant

Leaves
Leaves spread out to gather sunlight and carbon dioxide gas.

Stem
The stem holds the leaves up in the air.

Roots
Roots reach into the ground to soak up water.

From tiny to towering

There are about 400,000 different species, or types, of plants—in an enormous range of sizes—from nearly invisible **algae** to trees as tall as skyscrapers.

Grasses

Ferns

Flowers

Trees

I LOVE HIP HOP. HOW ABOUT YOU?

Plant mysteries

Unlike animals, plants do not hunt, sing, dance, fly, or growl, but that doesn't mean they can't do some things animals can do. Could plants be talking to each other right now, or even thinking? Do they like music? Can they grow in space? Could a plant swallow you alive? Read on to find out.

5

Why don't plants have mouths?

When we think about eating, we might think about munching on a sandwich or a bag of potato chips. You won't see a plant doing anything like that!*

Animals usually get the energy they need by eating food with their mouths.

Some animals, such as cats, will hunt other animals to eat.

WHAT ARE YOU LOOKING AT?

Other animals, such as cows, just munch on grass.

I'll make my own, thanks!
But for a plant, food is a totally different thing. Most of a plant's food comes from INSIDE the plant itself. It makes its own food through a process called **photosynthesis**, which means "making with light."

*Actually, there are some plants that eat meat, but they're not normal! Find out about them on page 24.

Photosynthesis mostly happens in leaves.

IN
- Sunshine
- Carbon dioxide gas
- Water from roots

OUT
Oxygen is released as waste or an unused gas. Food for the plant travels from the leaves to the rest of the plant.

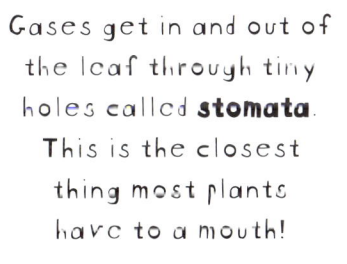

Gases get in and out of the leaf through tiny holes called **stomata**. This is the closest thing most plants have to a mouth!

Stomata

Using the food

The plant uses food to build new **cells** and plant parts, such as flowers, seeds, fruits, and more roots, stems, and leaves.

But what's this for?

YUM YUM!

Roots

Take your vitamins!

OK, that's not the ONLY way plants get food. They also take in chemicals, which are substances in the soil, through the water they soak up.

It's like the way our bodies need vitamins and minerals. They are essential to our health, but they don't make up most of what we eat. When you add plant food to the soil or a **nutrient** booster called fertilizer, it's like giving a plant its vitamins and minerals.

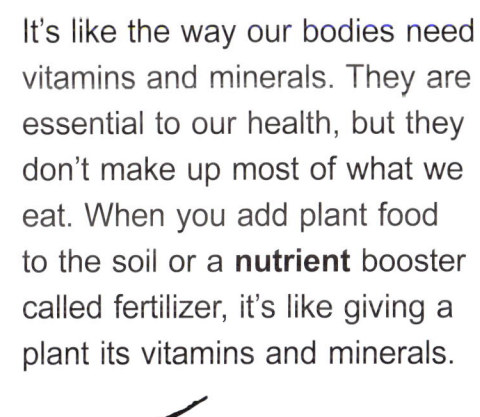

Could we exist without plants?

The answer to this is simple. It's a great big... NO!

If all the plants disappeared tomorrow, all the humans (and most other animals) would soon be gone, too.

Plants keep us alive in two main ways:

Food to eat (the food chain)

In the **food chain**, plants become food for some animals, which are then eaten by other animals.

Plants take in sunlight and grow.

Plant-eating animals eat the plants.

Meat-eating animals eat the plant-eating animals.

Air to breathe

Plants also give us oxygen, which living things need to breathe.

Without plants, carbon dioxide gas would build up in the air, and there wouldn't be enough oxygen to breathe.

① Plants take in carbon dioxide gas from the air and give out oxygen gas.

② Humans and animals do the opposite. They breathe in oxygen and breathe out carbon dioxide.

That's one reason why cutting down forests is bad for the environment. We need forests to absorb carbon dioxide and keep the air breathable.

8

Super useful

As if that isn't enough, there are so many other things plants provide, such as…

Crops, such as coffee

Building materials

Habitats for animals to live in

Materials, such as rubber

Medicines

SEED YOU LATER!

A secret mountain hideaway for seeds

Plants are SO important that we really can't afford to lose them. To make sure we don't, a huge seed bank has been built inside a mountain on a remote island in Norway. Seeds from thousands of different plants are stored there. If any plant species get wiped out by climate change or other disasters, we can use the seeds to start growing them again.

Entrance to the seed bank

Why can't plants walk around?

In the famous science-fiction story *The Day of the Triffids*, killer plants walk around on their stumpy, root-like feet.

In real life, plants don't walk — but why not?

COMING TO GET YOU!

These roots weren't made for walking!

Let's face it—plants don't have legs or feet. That makes it pretty difficult to walk! Most plants are also rooted into the soil, so they can't move.

This allows the roots to soak up water and helps keep the plant standing upright.

Plants can also get sunshine and air without going anywhere. So they don't really need to walk around.

Roots

Seeds on the move

So how do new plants grow in different spots? Like all living things, plants **reproduce**. They make seeds that can grow into new, young plants.

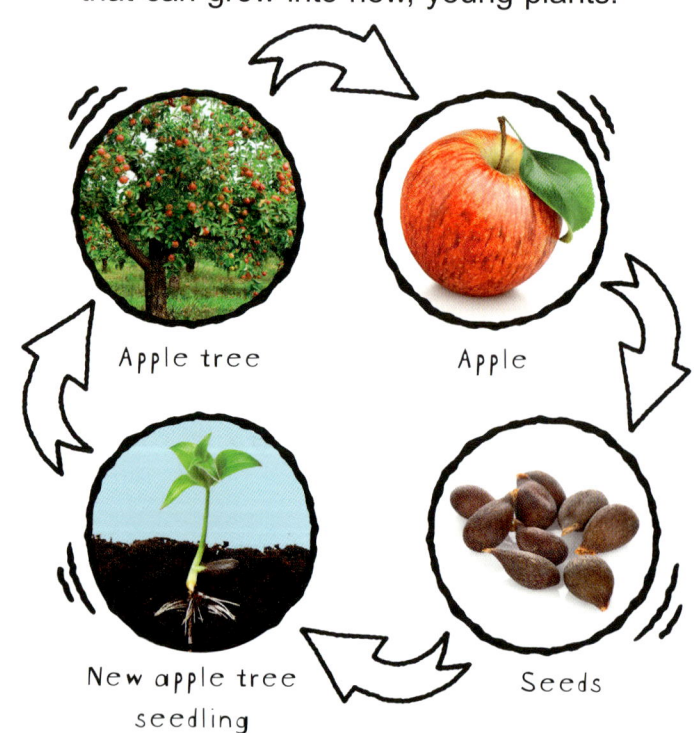

Apple tree → Apple → Seeds → New apple tree seedling

10

But if plants just dropped their seeds right next to them, they would end up all squashed into the same space fighting for **nutrients**. Instead, plants have amazing ways to make sure their seeds can travel as far as possible to find new places to grow:

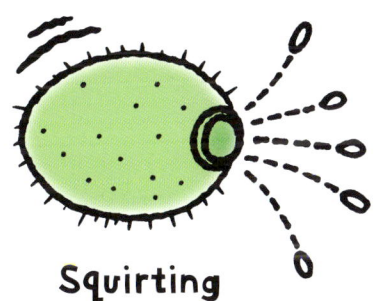

Squirting
Squirting cucumbers explode and shoot their seeds outward.

Through a stomach
Birds eat berries, then fly away and poo the seeds out in their droppings.

Hitching a ride
Some plants have burrs, or pouches that hold seeds. Burrs travel by hooking onto animal fur.

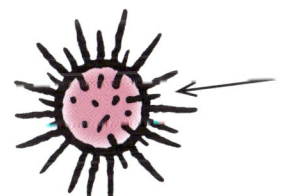

Floating
Some plants along coasts make seeds that can float across water.

Dandelion seed

Flying
Winged or fluffy seeds fly away on the wind.

Mary's bean

Which seed travels the farthest?

THE WINNER IS...
MARY'S BEANS!

These seeds, also called sea beans, can float 15,000 miles (24,000 km) across oceans. Mary's beans come from vines that are in the morning-glory family.

11

How do cacti survive in the desert?

Plants need water, and deserts are known for being VERY dry. But there must be some water somewhere, otherwise even a cactus couldn't survive.

Here's how they do it!

Catch the water

Most deserts do get some rain, but it's rare and doesn't last long. A cactus stores water to use until the next time it rains.

A cactus has a thick stem and branches with needles, called spines, instead of flat leaves.

Ribs expand so the cactus can swell with water.

Spongy material inside soaks up water.

Waxy, waterproof skin stops water from escaping out.

Shallow, wide roots spread far out and grab water as soon as it hits the ground.

A saguaro cactus in the Mexican desert

12

Get off my water!

Unsurprisingly, many desert animals would like to eat a cactus to get to the water.

That's why most cacti have spiky spines!

MMM, LOOKS JUICY... OUCH!

Drought-busting desert dwellers

The amazing baobab tree stores water in its trunk—as much as 31,700 gallons (120,000 liters), or enough for a small swimming pool!

The largest baobab's trunk grew to be 154 feet (47 m) wide.

The watery pulp inside also tastes awful and can make you sick.

Lithops or pebble plants are small desert plants that look amazingly like stones.

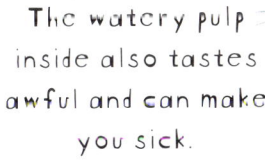

Spot the plant!

No rain... no plants!

Some places, such as parts of the Atacama Desert in Chile, have so little rain that they have no plants at all.

The monstrous-looking welwitschia sucks water from deep underground. It also collects droplets from fog. It grows incredibly slowly, but can live for over 1,000 years!

I'M OUTTA HERE!

Why do flowers smell nice?

It's very nice of flowers to fill our gardens with beautiful scents and colors, and make ingredients for our bubble baths, soaps, and perfumes.

But why do they do this?

Pollen-carriers

It's all because of the way plants reproduce. To make seeds, many plants need to exchange **pollen** cells with other plants in the same species. This process is called **pollination**.

 But, of course, plants can't walk around (see page 10), so they need help to move the pollen.

Wind pollination

Some plants, such as grass, just let their pollen blow away on the wind.

14

Animal pollination

Other plants use insects or other animals to carry their pollen from plant to plant. The plants make flowers that contain sugary **nectar** for insects to feed on. Sweet scents and bright colors attract the insects and animals, leading them where the plants need them to go.

THIS WAY, BEES!

Smell attracts insects from far away.

Up close, color and shape help insects find the flower.

This honeybee is dusted with pollen from the previous flower it visited.

Some of the pollen comes off as it visits other flowers, pollinating them.

No bees for me!

Bees are the most famous **pollinators**. Others include bats, birds, moths, butterflies, beetles wasps, lizards, and lemurs. They each have their favorite flowers, which they feed on and help to pollinate.

This hummingbird is pollinating a zinnia flower.

This flower smells awful!

STINKY MEAT – FANTASTIC!

Not all flowers smell nice—to us, at least. The rafflesia (which also happens to be the world's biggest flower) smells like rotting meat to attract its pollinators. Carrion flies sniff out dead animals to lay their eggs in, so the rafflesia smells great to them!

Rafflesia

15

How can a seed grow after thousands of years?

In 2005, scientists planted some date palm seeds that had been found in the ruins of an ancient fortress in Israel. The seeds were an amazing 2,000 years old.

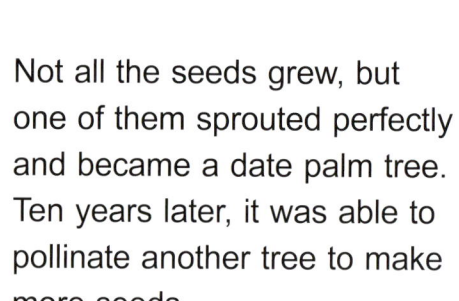

Date palm seeds

Not all the seeds grew, but one of them sprouted perfectly and became a date palm tree. Ten years later, it was able to pollinate another tree to make more seeds.

Great at waiting!

To **germinate**, or start growing into a new plant, a seed needs three main things:

sunlight... water... ...and a warm enough temperature

The tree was named Methuselah after an ancient man who was said to have lived to age 969. The seed was actually more than twice his age!

16

 # It may take some time to find all these things.

Seeds simply wait until the conditions are right.

Horse chestnut seeds wait in the soil until springtime before starting to grow.

Salsify seeds drift on the wind using their fluffy parachutes. They can travel for weeks or months before landing in a good place to grow.

Sand verbena seeds can wait for years in desert sands until a rain shower lets them start growing.

Seeds in a package from a garden center also have to wait until someone buys them and plants them.

Are seeds ALIVE?

YES—It may not look like it, but seeds are alive (unless they've been cooked, or damaged in some way) and can stay alive for a long time

However, a seed is **dormant**, or asleep, until it germinates. It's a bit like an animal **hibernating** in the winter, when its body slows down and it doesn't need to eat.

 But a seed slows down even more. In fact, it does nothing at all. Not all seeds will last 2,000 years before waking up, but it's definitely possible!

17

Do plants have feelings?

Plants don't have brains, so they can't think and feel like we do. They don't feel jealous that the plants next door get more sunshine, or get annoyed about the rain, or fall in love with each other. (As far as we know!)

However, plants can sense things, and respond. For example, if they sense light they grow toward it. But that's not all...plants feel all kinds of things!

Do plants mind being eaten?

Like animals, plants try to stay alive and prefer not to be eaten. That's why some have thorns or stinging leaves to keep hungry plant-eaters away.

Some plants can even tell when they're being eaten!

When caterpillars start munching on a rockcress plant, the plant makes bad-tasting chemicals. It spreads through the leaves leaving caterpillars with a nasty taste.

Rockcress

18

But it gets weirder...

Some scientists tried playing the rockcress plants a recording of the sound of munching caterpillars. Even without actually being eaten, the rockcress still made the bad-tasting chemicals! That means the plants could HEAR the caterpillars, and knew what the sound meant.

Do plants like music?

Gardeners sometimes claim their plants grow better if they play them music or sing to them. Experiments have shown that it's true! Music, and other sounds, often do seem to make plants grow better, but no one is sure why.

I felt that!

Plants may not feel emotions, but some can feel it when they are touched. The mimosa plant closes up and droops its leaves when you touch it, to try to avoid being eaten.

Climbing plants with tendrils, such as pea plants, reach out until they feel something to grab onto, then coil around it.

I'm coming to get you!

Scariest of all is the dodder plant. It's a **parasite** that winds around and lives on other plants, such as the tomato. When its favorite prey plants are nearby, the dodder can smell them, and moves toward them.

Dodder

19

Why are trees so big?

We're so used to trees that we don't often stop to think about how amazing they are. Trees are HUGE plants that tower over us. Some can grow taller than a 20-story building.

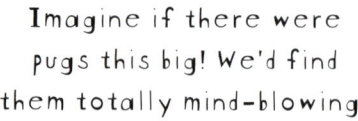

Imagine if there were pugs this big! We'd find them totally mind-blowing.

The biggest trees

Most trees are big compared to the average plant, but these enormous trees are record holders.

Tallest tree species: Coastal redwood

These **conifer** trees live on the Pacific coast of California. Many grow more than 300 feet (90 m) tall, and their trunks can measure more than 26 feet (8 m) across.

Tallest individual tree: Hyperion

The world's tallest trees are often given their own name! A tree named Hyperion is officially the tallest coastal redwood, at 380 feet (115 m) tall.

Biggest tree: General Sherman

It's not the tallest or the widest, but a giant sequoia named General Sherman, is the biggest and heaviest tree overall, thanks to its huge trunk. It is 275 feet (84 m) tall, and weighs 2.7 million pounds (1.2 million kg)—as much as 11 blue whales!

Hyperion 380 feet (115 m)

Leaning tower of Pisa 187 feet (57 m)

Statue of Liberty 151 feet (46 m)

Giraffe 19.7 feet (6 m)

11 OF ME? THAT'S RIDICULOUS!

Why do trees get so tall?

The first plants were all small. Trees **evolved** after plants developed hard, wooden stems, which gave them support to grow bigger.

Wattieza, one of the first trees, dates from about 380 million years ago.

Tall and strong

Being tall, with thick wooden trunks, helps trees survive in several different ways:

- It keeps their leaves out of the reach of many types of animals.
- It helps them reach higher than other plants to get more sunlight.
- Tall trees are better at surviving forest fires.

At home in a tree

You can even live in a tree! Some people do—in treehouses, or in tents on flat platforms in the branches.

Eco-activist Julia Butterfly Hill lived in a coastal redwood tree named Luna for more than two years to prevent it from being cut down.

Can plants talk to each other?

You won't see plants having a conversation and talking to each other in words. But they do have other ways of sending each other messages.

Plants have two main ways of "talking" to each other.

① Through the air

When a plant's leaves get munched or damaged, it releases chemicals into the air.

NICE WEATHER TODAY!

SHHH, THERE'S A HUMAN!

MMMMM, THE LOVELY SMELL OF CUT GRASS!

Think of the smell of freshly cut grass. That's the chemicals that escape when the blades of grass get cut off.

AAARRGGH! LAWNMOWER!

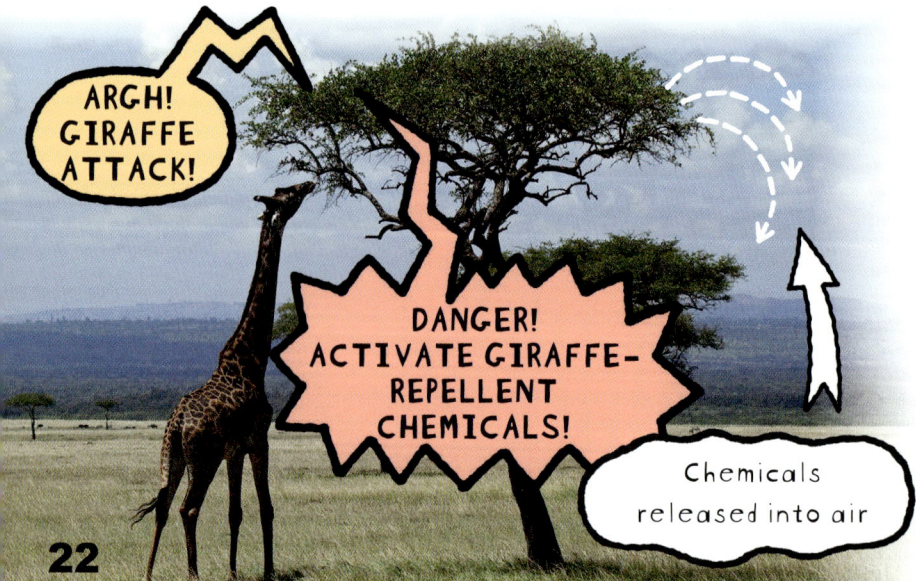

ARGH! GIRAFFE ATTACK!

DANGER! ACTIVATE GIRAFFE-REPELLENT CHEMICALS!

Chemicals released into air

Some plants can sense the chemicals released by another plant that is being eaten. They take action by making their own leaves taste bad before the animal even reaches them.

22

② Through the ground

In the soil, special types of **fungi** live around plant roots. This is good for both the plants and the fungi:

- The fungi help the plants soak up more water and nutrients.
- In return, the plants give the fungi food from their roots.

On the network

The fungi have **networks** of thin roots that spread through the soil. Scientists have found that these networks can actually link different plants together. Chemical signals can spread through the roots from one plant to another, to send messages, such as warning of an **aphid** attack.

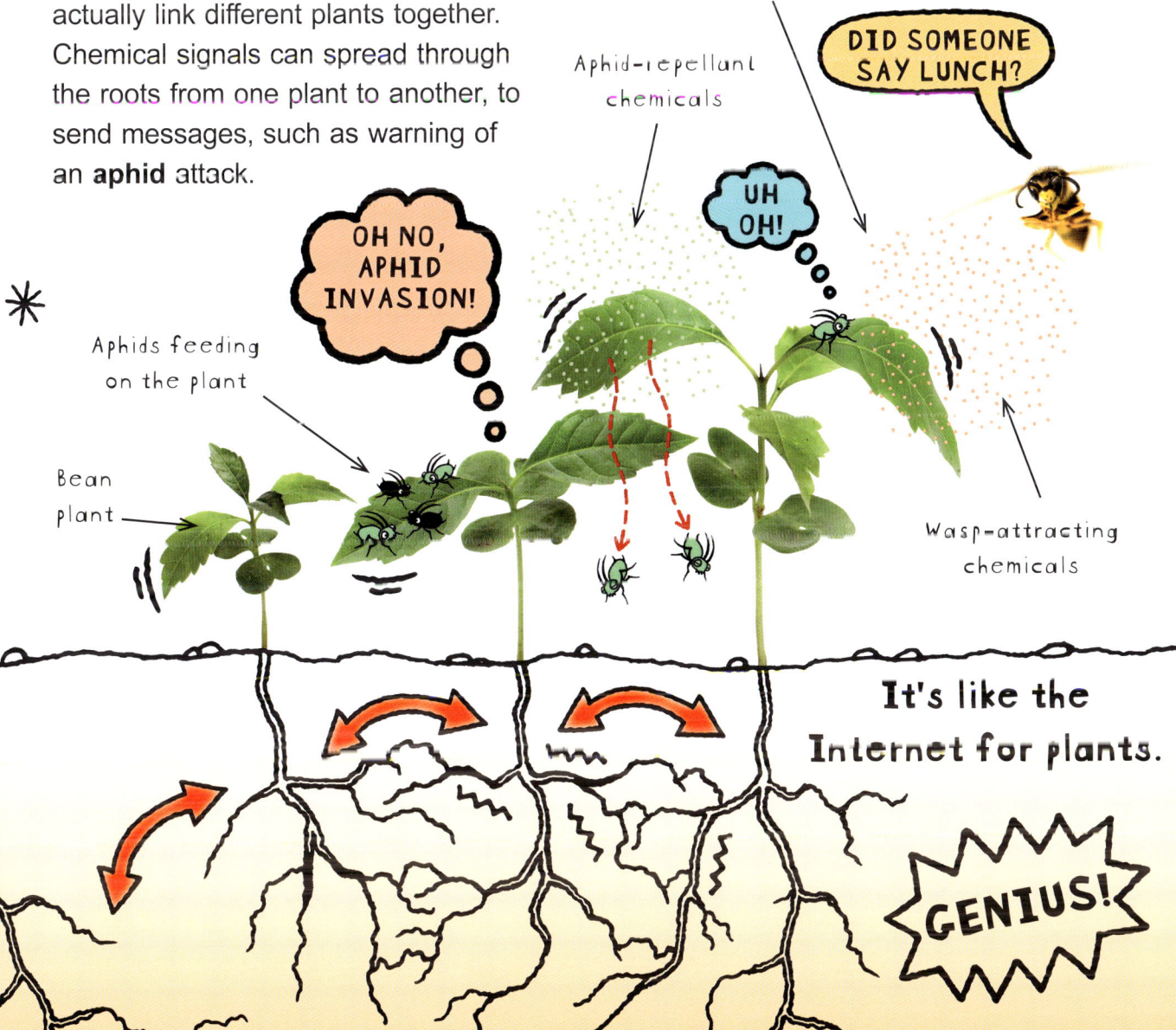

This bean plant also releases chemicals into the air to attract wasps. Why? Because wasps eat aphids!

Aphid-repellant chemicals

DID SOMEONE SAY LUNCH?

OH NO, APHID INVASION!

UH OH!

Aphids feeding on the plant

Bean plant

Wasp-attracting chemicals

It's like the Internet for plants.

GENIUS!

23

How can a plant eat a fly?

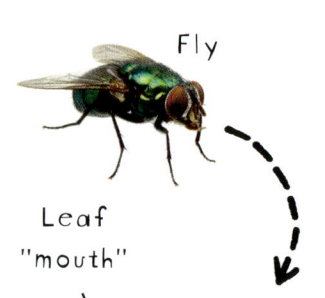

Fly

Leaf "mouth"

Venus fly trap

Strange as it may seem, some plants really do catch and eat animals. **Carnivorous plants** usually live in places where there aren't many nutrients in the soil. To get the nutrients they need, they top up their diet with meat!

MMMM, DELICIOUS...

SNAP!

Snap!

You may have heard of the famous Venus fly trap. It has mouth-like leaves with jaws and even teeth! The inside of each "mouth" is sensitive to touch, and when a fly lands inside it, it quickly closes, trapping its prey.

However, it's not really a mouth, and the plant doesn't swallow the fly. Instead, it releases juices that dissolve the fly, so that it can be soaked up by the plant as a liquid.

Giant Malaysian pitcher

At least the fly trap only eats flies. Pitcher plants can catch animals! A pitcher is a slippery jug-shaped leaf. The leaf is filled with liquid that dissolves any prey that falls in.

The giant Malaysian pitcher plant has the deepest pitchers, up to 16 inches (40 cm) tall. Unlucky creatures found inside them have included ants, centipedes, frogs, lizards, small birds, and even a RAT.

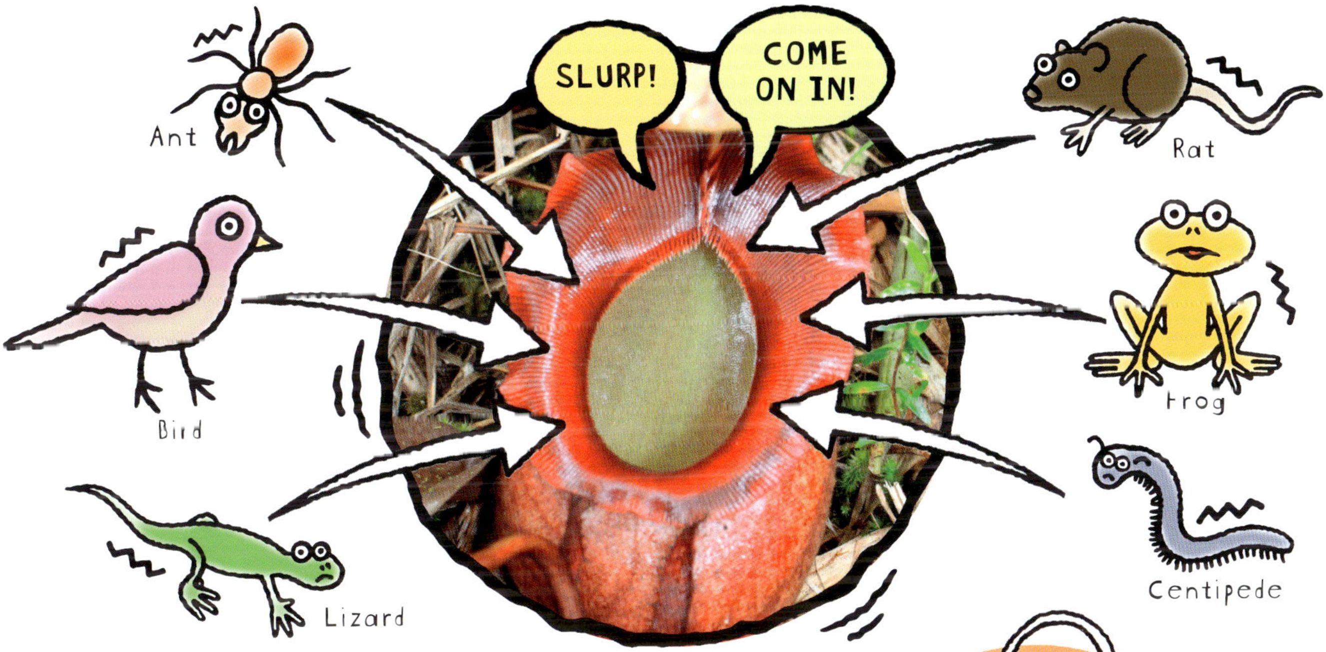

SLURP! COME ON IN!

Ant
Bird
Lizard
Rat
Frog
Centipede

Am I safe?

Are there any plants that could eat something bigger—perhaps even you? After all, there are people-eating plants in many folk tales and in science fiction.

The vampire vine from South America is said to wrap itself around animals or people and suck out all of their blood.

And a terrifying tree from Africa has long snake-like branches that grab hold of its human prey. Its name is Ya-Te-Veo, which means "I see you." Yikes!

Luckily for us, these murderous monsters are both totally made up. Phew!

Are there plants in space?

People often wonder if aliens exist, but we don't usually think of those aliens as plants.

Could plants be living on other planets?

So far, we haven't found life on any planet except our own. But that doesn't mean it's not there.

Light from the Sun

Like Earth, other planets and moons in the solar system orbit, or move around, the Sun. It's possible they could have plants, which take energy from sunlight.

Many of these worlds are icy cold, so the plants might be a bit like the tough, small plants that grow in Earth's chilly polar regions.

Plants could possibly grow on Jupiter's moon, Europa.

26

Other suns

Astronomers have discovered many other stars similar to our Sun, with their own planets orbiting them. Called **exoplanets**, some of these are similar to Earth. They have a warm climate, plenty of water, and light from their own star, so it's possible plants could grow there.

A jungle-covered exoplanet might look like this.

Purple plants

Plants on Earth contain a green substance called **chlorophyll**, which absorbs light for photosynthesis. But other chemicals can do this too, including a purple one called retinal. Some alien plants could be retinal-based, making them purple!

Alien plants would have evolved differently than Earth plants. They might look really strange, such as these!

Plants in space

We know there definitely are SOME plants in space, because we sent them up there on space missions. Astronauts on the International Space Station have grown several plants to see how very little gravity affects such plants as lentils, peas, sunflowers, and zinnias.

WE'RE OUT OF THIS WORLD!

27

Quick-fire questions

Why don't plants freeze to death in winter?

Some plants do die in freezing weather, especially tropical plants because they have not evolved to cope with the cold. Others can survive icy weather by making chemicals that stop water from freezing solid, which would damage their cells.

Can plants disguise themselves?

A few can. The pebble plant (see page 13) looks like a rock to avoid being eaten. The dead nettle can't sting, but it looks just like a real stinging nettle to scare off hungry animals. The bee orchid flower looks just like a particular type of female bee to attract male bees to pollinate it.

Which plant has the deadliest poison?

Scientists disagree about which is the most poisonous plant of all. The deadliest include aconite, also known as monkshood, oleander, and the castor bean plant, which contains the killer poison ricin.

COME ON OVER BEES!

Bee orchid

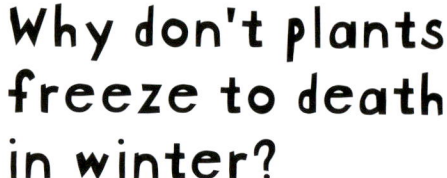

Castor beans

28

Which plant has the worst sting?

Don't go near a gympie-gympie! This plant from Australia and southeast Asia is covered in tiny stinging hairs that get stuck in your skin if you touch it. The pain is agonizing, and can last for years after you've been stung.

Is a mushroom a plant?

No! Mushrooms belong to the fungi family, along with molds, yeast, and the fungi that grow around plants' roots. Unlike plants, fungi don't get energy from sunlight. They take in food from soil, plants, or animals. Athlete's foot is an example of a fungus that can grow between your toes. **Eeewww.**

Why are plant roots hairy?

Many plants have small hair-like parts on their roots. These tiny roots increase the surface area of the larger roots, meaning they can suck up more water. Root hairs can also reach into small gaps in the soil to find more water.

Which plant grows the fastest?

Bamboo, which is a type of giant grass, is the fastest-growing of all plants. Some species can grow 35.5 inches (90 cm) in a single day!

LUCKY ME—I LOVE BAMBOO!

29

Glossary

algae Plant-like living things, that live in water and use the Sun's energy to make food

aphid A small, sap-sucking insect

carbon dioxide A gas found in the air and used by plants to make their food

carnivorous plant A plant that catches and digests small animals, such as insects

cells The tiny building blocks that all living things are made up of

chlorophyll A green chemical inside plants, that takes in sunlight and helps plants make food

climate change Changes in weather patterns around the world, including increasing temperatures, or global warming

conifer A tree with cones and leaves shaped like needles

dormant Inactive or "asleep" for a long time, but still alive

evolved To have changed over a long period of time

exoplanet A planet outside our solar system

food chain A pattern of eating and being eaten, beginning with plants and ending with large animals

fungi A species, such as mold and mushrooms, that gets their food from both living and dead things

germinate To grow from a seed to a plant

habitats The natural homes of living things, such as a forest or coral reef

hibernate To spend the winter in an inactive state, similar to a long sleep

nectar Sweet liquid made by plants as a food to attract pollinators

networks Groups or systems of connected parts

nutrient A substance that living things get from food needed for growth and good health

oxygen A gas in air that humans, animals, and plants need to breathe

parasite A plant that lives on or inside a bigger plant and gets its food from it

photosynthesis The process by which plants use sunlight to combine carbon dioxide and water to create food

pollen A powder in flowers that is needed to make new flowers

pollination The process of transferring pollen from one flower to another to allow plants to reproduce

pollinators Animals, such as honeybees or hummingbirds, that pollinate plants

prey A plant or animal that is trapped or hunted by another plant or animal for food

reproduce To have babies, or make more living things of the same species

stomata Tiny holes on the surface of leaves that let gases in and out

Learning More

Books

Claybourne, Anna. *Killer Plants and Other Green Gunk*. Crabtree Publishing, 2015.

Hickman, Pamela. *Nature All Around: Plants*. Kids Can Press, 2020.

Hudd, Emily. *How Long Does a Redwood Tree Live?* Capstone Press, 2019.

Machajewski, Sarah. *Meat-Eating Plants*. Gareth Stevens, 2019.

Websites

www.dkfindout.com/us/animals-and-nature/plants
This website has lots of useful plant facts, photos, and quizzes.

www.coolkidfacts.com/germination-for-kids
Visit this site to learn all about seed germination.

www.scienceforkidsclub.com/carnivorous-plants.html
Learn more about carnivorous plants here.

www.sciencebuddies.org/science-fair-projects/project-ideas/plant-biology
Visit this site for dozens of amazing plant experiments and science projects.

Every effort has been made by the Publishers to ensure that the websites in this book are suitable for children, that they are of the highest educational value, and that they contain no inappropriate or offensive material. However, because of the nature of the Internet, it is impossible to guarantee that the contents of these sites will not be altered. We strongly advise that Internet access is supervised by a responsible adult.

Index

A
algae 5

B
bamboo 29
baobabs 13
bee orchids 28

C
cacti 12–13
carbon dioxide 4–5, 7, 8
carnivorous plants 6, 24–25
chlorophyll 27
coastal redwoods 20–21
communication 5, 22–23
conifers 20–21

D
date palms 16
deserts 12–13, 17
dodders 19

F
flowers 5, 7, 14–15, 28
food chain 8–9
fungi 23, 29

G
germination 16–17
Giant Malaysian pitcher plant 25
giant sequoia 20
grass 5, 6, 14, 22, 29
gympie-gympie 29

H
Hill, Julia Butterfly 21

I
insects 15, 18–19, 23, 24, 25, 28

L
leaves 5, 7, 12, 18–19, 21, 22, 24–25
lithops 13, 28

M
mimosa 19
music 5, 19

N
nectar 15
nutrients 7, 23, 24, 25

O
oxygen 7, 8

P
pea plants 19, 27
photosynthesis 6–7, 27
planets 26–27
poison 28
pollination 14–15, 16, 28

R
rafflesia 15
reproduction 10, 14, 15
rockcress plants 18, 19
roots 5, 7, 10, 12, 23, 29

S
scent 14–15
seed bank 9
seeds 7, 9, 10, 11, 14, 16–17,
space 5, 26–27
stems 5, 7, 12, 21
stomata 7
sunlight 4–5, 7, 8, 10, 16, 18, 21, 26–27, 29

T
trees 5, 16, 20–21

V
Venus fly trap 25–25

W
water 4–5, 7, 10, 12, 13, 16, 23, 27, 29
wattieza 21
welwitschia 13

32